EMMANUEL JOSEPH

Access to Clean Water in Developing Countries

Contents

1

Access to Clean Water in Developing Countries

1.1 The Global Water Crisis

Water is one of the Earth's most precious resources, essential for life, health, and well-being. Yet, despite its undeniable importance, access to clean and safe drinking water remains a significant challenge in many parts of the world, especially in developing countries. This introductory chapter sets the stage for our exploration of this pressing issue by highlighting the overarching global water crisis.

1.2 The Importance of Clean Water

The fundamental importance of clean water cannot be overstated. Water is not only essential for hydration but also plays a critical role in sanitation, agriculture, and industry. Lack of access to clean water can lead to waterborne diseases, malnutrition, economic stagnation, and a host of other problems. Understanding the gravity of this issue is the first step toward addressing it effectively.

1.3 Purpose and Scope of the Book

In this book, we will delve into the ongoing challenges faced by billions of people in developing countries in their quest for clean and safe drinking water. We will examine the multidimensional aspects of this problem, from the sources of water contamination to the health impacts of inadequate access. The book's primary objectives are as follows:

1.3.1 Raising Awareness

This book aims to raise awareness about the global water crisis and its disproportionate impact on developing countries. By presenting facts, figures, and real-life stories, we hope to emphasize the urgency of the situation.

1.3.2 Analyzing the Root Causes

We will explore the root causes of inadequate access to clean water, such as inadequate infrastructure, water pollution, economic constraints, and social disparities. This analysis will provide a comprehensive understanding of the complexities involved.

1.3.3 Showcasing Solutions

While the problem is daunting, it is not insurmountable. Throughout this book, we will showcase various initiatives, both governmental and non-governmental, that have made strides in improving access to clean water. We will also discuss technological innovations and community-led solutions that offer hope for a better future.

1.3.4 Encouraging Action

Ultimately, this book is a call to action. We hope to inspire individuals, organizations, and governments to take steps, both big and small, to address the water crisis. Clean water is a fundamental human right, and it is within our reach to ensure that everyone has access to it.

1.4 Structure of the Book

To provide a comprehensive understanding of the issue of access to clean water in developing countries, this book is organized into 12 chapters, each addressing specific aspects of the problem. Here is an overview of what to expect in the following chapters:

Chapter 2: The Water Crisis in Developing Countries - A detailed examination of the current state of water access in these nations, highlighting regional disparities and challenges.

Chapter 3: Waterborne Diseases and Health Impact - Exploring the devastating link between contaminated water and diseases, with case studies illustrating the human cost.

Chapter 4: Water Sources in Developing Countries - An exploration of the types of water sources used and the issues related to surface water, groundwater, and unimproved sources.

Chapter 5: Water Quality and Contamination - A look at common contaminants and pollutants, their sources, and their impact on water quality and health.

Chapter 6: Infrastructure and Access Challenges - Analyzing the obstacles to building and maintaining clean water infrastructure and the barriers faced by individuals in accessing clean water.

Chapter 7: Governmental and Non-Governmental Initiatives - An examination of the roles governments and non-governmental organizations play in addressing water issues, with success stories illustrating their efforts.

Chapter 8: International Aid and Development Assistance - A discussion of the contributions of international organizations and donor countries, the challenges they face, and the effectiveness of their aid programs.

Chapter 9: Community-Led Solutions - An exploration of the importance of community involvement in solving water access issues, including examples of successful grassroots initiatives.

Chapter 10: Technological Innovations for Water Access - Highlighting emerging technologies and low-cost solutions that hold promise for improving water access in developing countries.

Chapter 11: Economic and Social Implications - Examining the economic impact of clean water access, social and gender disparities, and the recognition of water as a fundamental human right.

Chapter 12: Future Prospects and Recommendations - Discussing the path forward to address the water crisis, offering policy recommendations, and emphasizing the role of individuals in creating positive change.

As we journey through these chapters, we will gain a deeper understanding of the ongoing challenges and potential solutions related to access to clean water in developing countries. Together, we can work towards a future where clean water is not a luxury but a fundamental right for all.

2

Chapter 2: The Water Crisis in Developing Countries

2.1 Introduction

Access to clean water in developing countries is a critical issue that affects the well-being, health, and development of millions of people. In this chapter, we will delve into the current state of the water crisis in these regions, shedding light on the challenges they face and the disparities that exist.

2.2 The Current State of Water Access

In many developing countries, access to clean and safe drinking water is far from guaranteed. The availability of clean water varies widely, both within and between nations. Some key observations include:

- Regional Disparities: Rural areas often have less access to clean water than urban areas. Additionally, water access can vary significantly between regions, with some areas suffering from chronic shortages.

- Population Growth: As populations in these countries grow, the demand for water increases. In many cases, infrastructure and resources are insufficient

5

to keep pace with this demand.

- Climate Change Impacts: Climate change has brought about increased droughts and erratic weather patterns, affecting water sources and exacerbating water scarcity issues.

2.3 Challenges in Water Access

Several challenges contribute to the water crisis in developing countries:

- Infrastructure Deficiency: Many regions lack adequate infrastructure for water supply and sanitation. Piped water networks, treatment facilities, and distribution systems are often inadequate or non-existent.

- Contamination and Pollution: Water sources are frequently contaminated with pollutants, pathogens, and chemicals. This contamination leads to waterborne diseases and health issues.

- Economic Constraints: High costs associated with building and maintaining water infrastructure can strain the budgets of developing nations. This can lead to deferred maintenance and insufficient investment.

2.4 Regional Disparities and Case Studies

To illustrate the severity of the issue, we'll examine specific regions and case studies. These examples will highlight the water access challenges faced by people in different parts of the world. Some examples may include:

- Sub-Saharan Africa: This region faces significant water access challenges, with many communities relying on unsafe water sources, leading to high disease burdens.

- South Asia: High population density and limited access to clean water in some areas contribute to severe water-related problems.

- Latin America: Issues with water access, sanitation, and contamination are present in several Latin American countries.

2.5 Impact on Daily Life

The lack of access to clean water has a profound impact on the daily lives of individuals in developing countries. This impact extends beyond the obvious challenge of obtaining safe drinking water. People face difficulties in sanitation, hygiene, and livelihoods. Health is compromised, and the burden of water collection and disease management falls heavily on women and children.

2.6 Conclusion

The global water crisis, particularly in developing countries, remains a daunting challenge with far-reaching consequences. The disparities in water access and the obstacles faced by individuals in these regions underscore the urgent need for concerted efforts to address this issue. In the following chapters, we will explore the health impact of waterborne diseases, the sources of water contamination, and efforts to improve water infrastructure and access, providing a comprehensive view of the ongoing challenges and potential solutions in the quest for clean and safe drinking water.

3

Chapter 3: Waterborne Diseases and Health Impact

3.1 Introduction

In this chapter, we delve into the devastating link between contaminated water and waterborne diseases, shedding light on the profound health impact of inadequate access to clean water in developing countries. Waterborne diseases represent a significant threat to the well-being and livelihoods of millions of individuals in these regions.

3.2 The Health Toll of Contaminated Water

Contaminated water is a breeding ground for a wide range of waterborne diseases, including cholera, dysentery, typhoid, and various gastrointestinal infections. These diseases can cause severe symptoms, lead to long-term health complications, and even be fatal, particularly in children and the elderly.

3.3 Vulnerable Populations

Children are among the most vulnerable to waterborne diseases. Inadequate access to clean water can lead to high rates of infant mortality and stunted growth. Moreover, women, who often bear the primary responsibility for water collection and sanitation, are at greater risk of infection due to their frequent exposure to contaminated water sources.

3.4 Case Studies

This chapter will present case studies to illustrate the real-world impact of waterborne diseases:

- Cholera Outbreaks in Haiti: The 2010 cholera outbreak in Haiti, which resulted from contaminated water sources, serves as a poignant example of the devastating consequences of waterborne diseases.

- Impact on Child Mortality in Sub-Saharan Africa: We will explore how waterborne diseases contribute to child mortality in Sub-Saharan Africa, with a focus on the ongoing struggle to combat these diseases.

3.5 Preventive Measures and Treatment

While the health impact of waterborne diseases is severe, preventive measures and treatment strategies are available. We will discuss:

- Water Purification Techniques: Various methods, such as filtration, chlorination, and solar disinfection, can be used to make contaminated water safe for consumption.

- Sanitation and Hygiene Practices: The importance of proper sanitation and hygiene practices, including handwashing, to prevent waterborne diseases.

- Oral Rehydration Therapy (ORT): This simple and cost-effective treatment can save lives by rehydrating individuals suffering from severe diarrhea.

3.6 The Interplay with Malnutrition and Poverty

Waterborne diseases are often closely intertwined with malnutrition and poverty. The weakened state of individuals suffering from these diseases can exacerbate malnutrition, creating a vicious cycle of poor health and economic instability.

3.7 Conclusion

Access to clean water is not merely a matter of convenience; it is a fundamental human right with profound implications for health. Waterborne diseases continue to exact a heavy toll on the people of developing countries, particularly on children and vulnerable populations. In the chapters to come, we will explore the sources of water contamination, the challenges faced in securing clean water sources, and the efforts made to mitigate these health risks. Understanding the relationship between contaminated water and disease is crucial to addressing the global water crisis and improving the well-being of millions of individuals in need.

4

Chapter 4: Water Sources in Developing Countries

4.1 Introduction

Access to clean water begins with understanding the various water sources that people in developing countries rely on. This chapter explores the types of water sources commonly used in these regions and the challenges associated with them.

4.2 Categories of Water Sources

In developing countries, water sources can be categorized into three main types:

- Surface Water: This includes rivers, lakes, and streams. Surface water sources are readily available, but they are often exposed to contamination from human and animal activities.

- Groundwater: Water obtained from underground aquifers and wells. Groundwater is generally less vulnerable to contamination, but overextraction can lead to depletion and intrusion of pollutants.

- Unimproved Sources: These are sources that do not meet basic sanitary and safety criteria. They include unprotected wells, springs, and water from rivers or ponds without treatment.

4.3 Challenges with Surface Water

Surface water sources are commonly used, especially in rural areas. However, they come with several challenges:

- Contamination Risk: Surface water is susceptible to contamination from agricultural runoff, industrial discharges, and improper disposal of human waste.

- Seasonal Variability: These sources can be subject to seasonal variations in availability, which can lead to water scarcity during dry periods.

4.4 Groundwater as a Source

Groundwater is often considered a safer source of drinking water due to its natural filtration through the earth. However, it has its own set of challenges:

- Depletion: Over-extraction of groundwater can lead to the depletion of aquifers, causing water scarcity.

- Arsenic and Fluoride: In some regions, naturally occurring contaminants like arsenic and fluoride can be present in groundwater, posing health risks.

4.5 Unimproved Sources and Their Risks

Unimproved water sources, like unprotected wells and springs, are still commonly used, particularly in remote areas. These sources pose significant health risks due to their vulnerability to contamination.

4.6 Dependency on Multiple Sources

Many communities in developing countries rely on multiple sources for their water needs. This dependence on a mix of surface water, groundwater, and unimproved sources further complicates the challenge of ensuring safe and reliable access to clean water.

4.7 Conclusion

Understanding the types of water sources used in developing countries is essential for addressing the water crisis. Different sources come with varying levels of contamination risk, availability, and suitability for drinking water. In the following chapters, we will delve deeper into the issues of water quality and contamination, and explore solutions to provide safe and sustainable access to clean water. Addressing the challenges associated with these diverse water sources is a critical step in improving the well-being of communities in need.

5

Chapter 5: Water Quality and Contamination

5.1 Introduction

Access to clean water is not solely about the source; it's equally important to ensure the water's quality. In this chapter, we explore the common contaminants and pollutants that compromise water quality in developing countries, their sources, and their impact on health and well-being.

5.2 Common Contaminants

Water sources in developing countries often contain a variety of contaminants, including:

- Pathogens: Bacteria, viruses, and parasites from human and animal waste can lead to waterborne diseases such as cholera, dysentery, and typhoid.

- Chemicals: Agricultural runoff, industrial discharges, and natural minerals can introduce harmful chemicals like pesticides, heavy metals, and arsenic into water sources.

- Physical Impurities: Sediments, debris, and other physical impurities can affect water quality.

5.3 Sources of Contamination

Contamination can originate from various sources:

- Lack of Sanitation: Poor sanitation practices, open defecation, and improper disposal of waste can introduce pathogens into water sources.

- Agricultural Runoff: The use of fertilizers and pesticides in agriculture can contaminate water with chemicals harmful to human health.

- Industrial Pollution: Industrial activities can release a range of pollutants, including heavy metals and toxic chemicals, into water bodies.

5.4 Impact on Water Quality and Health

Contaminated water has a profound impact on health and well-being:

- Waterborne Diseases: Pathogens in water can lead to debilitating and potentially fatal diseases. Children are particularly vulnerable.

- Chronic Health Conditions: Long-term exposure to certain contaminants, like arsenic and fluoride, can result in chronic health conditions, such as skeletal fluorosis and skin lesions.

5.5 Vulnerable Populations

Certain populations are at higher risk of the health impacts of water contamination, including pregnant women, children, and those with compromised immune systems.

5.6 Monitoring and Water Quality Testing

To address water quality issues, regular monitoring and water quality testing are essential. These measures can help identify contamination sources and ensure the safety of drinking water.

5.7 Conclusion

Access to water alone is not sufficient; it must be clean and safe for consumption. Contamination of water sources in developing countries poses a significant health risk, leading to waterborne diseases and chronic health conditions. In the following chapters, we will explore the challenges of water infrastructure and access, governmental and non-governmental initiatives to improve water quality, and emerging technologies and innovations for purifying water. Understanding the sources and impact of water contamination is critical to addressing the global water crisis and ensuring that individuals have access to clean and safe drinking water.

6

Chapter 6: Infrastructure and Access Challenges

6.1 Introduction

The availability of clean water is not only about the source's quality; it's equally dependent on the infrastructure and distribution networks that transport water to communities. In this chapter, we examine the challenges and limitations related to water infrastructure and access in developing countries.

6.2 Inadequate Infrastructure

One of the primary challenges is the lack of proper infrastructure for water supply and distribution. This encompasses:

- Piped Water Networks: Many communities lack piped water systems that can deliver clean water directly to households, requiring residents to travel long distances to collect water.

- Treatment Facilities: Proper water treatment facilities are often lacking, leading to the distribution of untreated or poorly treated water.

- Storage Facilities: Insufficient storage capacity can result in water scarcity during dry periods or equipment breakdowns.

6.3 Barriers to Access

Access to clean water faces several barriers in developing countries:

- Distance: The time and effort required to collect water can be substantial, particularly for those living in rural areas. This places a heavy burden, especially on women and children.

- Economic Constraints: The costs associated with accessing water, including purchasing water from vendors or the installation of household water treatment systems, can be prohibitive for low-income families.

- Technological Limitations: Communities may lack access to or knowledge about water treatment technologies and safe storage solutions.

6.4 Impact on Health and Quality of Life

Inadequate infrastructure and limited access to clean water have severe implications for health and quality of life. The burden of water collection, often carried out by women and children, affects education, income, and overall well-being.

6.5 Climate Change and Infrastructure Vulnerability

Climate change exacerbates the challenges related to water infrastructure and access. More frequent and severe weather events can damage infrastructure and disrupt water supply systems, leaving communities vulnerable.

6.6 Local Solutions and Innovations

Communities and organizations are developing localized solutions to improve water access:

- Rainwater Harvesting: The collection and storage of rainwater for household use, particularly in arid regions.

- Community-Led Water Management: Empowering local communities to manage and maintain their water systems.

- Low-Cost Water Treatment Technologies: Affordable solutions that can purify water at the household level.

6.7 Conclusion

Addressing water infrastructure and access challenges in developing countries is vital to ensuring reliable and sustainable access to clean water. In the following chapters, we will explore governmental and non-governmental initiatives to improve infrastructure, international aid and development assistance, community-led solutions, and technological innovations that hold promise for a better water future. Understanding the complexities surrounding water infrastructure and access is crucial for tackling the global water crisis and improving the quality of life for millions of individuals in need.

7

Chapter 7: Governmental and Non-Governmental Initiatives

7.1 Introduction

The provision of clean and safe drinking water is a shared responsibility that often involves both government agencies and non-governmental organizations (NGOs). In this chapter, we explore the roles these entities play in addressing the water crisis in developing countries.

7.2 Governmental Efforts

Governmental agencies at the local, regional, and national levels are pivotal in addressing the water crisis:

- Water Policy and Regulation: Governments establish policies and regulations to ensure the quality, safety, and equitable distribution of water.

- Infrastructure Development: Governments invest in the construction and maintenance of water supply and distribution systems, including treatment plants and pipelines.

- Subsidies and Affordability Programs: To make clean water accessible to low-income populations, governments may provide subsidies or implement affordability programs.

7.3 Non-Governmental Organizations (NGOs)

NGOs play a vital role in addressing water-related challenges in developing countries:

- Service Provision: Many NGOs directly provide water services by building infrastructure, distributing water, and facilitating community-based projects.

- Advocacy and Awareness: NGOs raise awareness about the water crisis and advocate for policy changes and increased funding for water-related initiatives.

- Capacity Building: They build local capacity by training community members and local organizations to manage water resources and infrastructure.

7.4 Success Stories

This chapter will highlight examples of successful governmental and NGO initiatives that have made a significant impact on improving water access and quality in developing countries. These success stories can serve as models for effective approaches.

- India's Swachh Bharat Mission: India's massive campaign to improve sanitation and access to clean water in rural areas.

- Water.org's WaterCredit Program: An innovative initiative that leverages microfinance to provide loans for water and sanitation improvements.

7.5 Challenges and Limitations

Both governmental and non-governmental efforts face challenges such as funding limitations, bureaucratic obstacles, and the need for long-term sustainability.

7.6 International Collaborations

International collaboration, often involving donor countries and organizations, is essential to providing the necessary resources and expertise to tackle the global water crisis.

7.7 Conclusion

Governments and NGOs play a crucial role in improving access to clean water in developing countries. Their combined efforts, through policy development, infrastructure investment, service provision, advocacy, and capacity building, are fundamental to addressing this pressing issue. In the following chapters, we will explore the role of international aid and development assistance, community-led solutions, technological innovations, and the economic and social implications of water access. By understanding the various initiatives and partnerships in place, we can work towards a world where clean water is accessible to all.

8

Chapter 8: International Aid and Development Assistance

8.1 Introduction

International aid and development assistance play a significant role in addressing the global water crisis. In this chapter, we explore the involvement of international organizations, donor countries, and the impact of their contributions on improving access to clean water in developing countries.

8.2 The Role of International Organizations

Several international organizations, such as the United Nations, World Bank, and UNICEF, are actively engaged in water-related projects. Their roles include:

- Funding and Financing: Providing financial support to developing countries for water infrastructure development, sanitation, and capacity building.

- Policy and Advocacy: Advocating for improved water management policies, water-related sustainability goals, and raising awareness about water issues.

- Capacity Building: Supporting developing countries in building the capacity to manage and maintain water resources and infrastructure.

8.3 Donor Countries and Bilateral Aid

Donor countries, often part of bilateral agreements, provide direct aid to developing countries for water-related projects. They may offer grants, loans, or technical assistance to help build or upgrade water infrastructure and improve access to clean water.

8.4 Challenges and Effectiveness

International aid and development assistance face challenges and criticisms, including concerns about the effectiveness and sustainability of projects. These issues are explored in this chapter, along with discussions on accountability and transparency.

8.5 Case Studies

This chapter will highlight specific case studies that demonstrate the impact of international aid and development assistance in improving water access. These case studies may include examples from countries that have seen significant improvements in their water infrastructure and quality.

8.6 Future Directions

In this section, we discuss the future of international aid and development assistance in the context of addressing the global water crisis. This includes potential shifts in priorities, innovative funding mechanisms, and collaborative efforts to achieve sustainable development goals.

8.7 Conclusion

International aid and development assistance have a critical role in improving access to clean water in developing countries. The contributions of international organizations and donor countries, when well-managed and aligned with local needs, can make a significant impact on water infrastructure, sanitation, and overall well-being. In the following chapters, we will explore community-led solutions, technological innovations, and the economic and social implications of water access. By understanding the contributions and challenges associated with international aid, we can work toward a world where clean water is accessible to all.

9

Chapter 9: Community-Led Solutions

9.1 Introduction

Community involvement is a vital aspect of addressing the global water crisis. In this chapter, we explore the significance of community-led solutions and initiatives that empower local residents to take an active role in improving access to clean water.

9.2 The Importance of Community Involvement

Communities play a central role in solving water access challenges for several reasons:

- Local Knowledge: Residents have a deep understanding of their region's water sources, needs, and challenges.

- Ownership and Sustainability: When communities take ownership of water projects, they are more likely to ensure the long-term sustainability of the infrastructure.

- Empowerment: Engaging communities empowers them to actively participate in decision-making processes, leading to more effective and tailored solutions.

9.3 Examples of Community-Led Solutions

This chapter will provide examples of community-led initiatives and projects that have successfully improved water access in developing countries. These may include:

- Community-Managed Water Committees: Empowering local committees to manage and maintain water sources and infrastructure.

- Rainwater Harvesting: Communities working together to collect and store rainwater for domestic use.

- WASH Programs (Water, Sanitation, and Hygiene): Community-led efforts to improve sanitation and hygiene practices in tandem with clean water initiatives.

9.4 Challenges and Lessons Learned

Community-led solutions are not without challenges. This chapter will discuss common obstacles, such as financial limitations, training and capacity building, and the need for external support. It will also highlight the lessons learned from successful projects.

9.5 The Role of NGOs and International Organizations

Many non-governmental organizations and international agencies support community-led initiatives by providing funding, technical assistance, and expertise. This chapter will explore how these organizations collaborate with communities to enhance their water access efforts.

9.6 Sustainability and Scaling Up

Sustainability is a key consideration for community-led solutions. This section will delve into the strategies and mechanisms that help ensure the long-term viability of these projects. It will also discuss the potential for scaling up successful initiatives to benefit larger populations.

9.7 Conclusion

Community-led solutions are a fundamental part of addressing the global water crisis. They empower local communities to take control of their water resources and infrastructure, leading to more sustainable and effective solutions. In the following chapters, we will explore technological innovations, the economic and social implications of water access, and the path forward to secure clean water as a fundamental human right. By understanding the role of communities in addressing water access challenges, we can work towards a future where clean water is accessible to all.

10

Chapter 10: Technological Innovations for Water Access

10.1 Introduction

Technological innovations have the potential to play a transformative role in improving access to clean water in developing countries. In this chapter, we explore emerging technologies and low-cost solutions that hold promise for purifying water, enhancing distribution, and addressing water access challenges.

10.2 Emerging Technologies for Water Purification

Cutting-edge technologies are making it possible to purify water more efficiently and affordably:

- Solar Water Disinfection (SODIS): Using sunlight to disinfect water in PET bottles.

- Nanotechnology: Utilizing nanoparticles to remove contaminants and pathogens from water.

- Membrane Filtration: Advanced membrane systems to remove impurities from water.

10.3 Low-Cost Solutions

Inexpensive and locally available solutions have the potential to make a significant impact on water access:

- Ceramic Filters: Low-cost ceramic filters that can remove impurities from water.

- Biosand Filters: Filters using layers of sand and biological processes to treat water.

- Simple Hand-Pump Systems: Mechanical systems that allow communities to access groundwater more easily.

10.4 Challenges and Adoption Hurdles

While innovative technologies offer promise, they also face challenges in terms of scalability, cultural acceptance, and maintenance.

10.5 The Role of Innovation Hubs and Research

Innovation hubs and research institutions play a pivotal role in developing and testing new technologies for water access. These institutions collaborate with communities and organizations to pilot and scale innovative solutions.

10.6 Prospects for the Future

The future of technological innovations in the water sector holds exciting possibilities, from smart sensors for monitoring water quality to innovative water treatment methods. This section discusses the prospects for the future and potential game-changing technologies.

10.7 Conclusion

Technological innovations are revolutionizing water access in developing countries. From low-cost, community-based solutions to advanced purification methods, these innovations have the potential to provide clean water to those who need it most. In the following chapters, we will explore the economic and social implications of water access, as well as the path forward to ensure that clean water becomes a fundamental human right for all. Understanding the potential of technology is critical to addressing the global water crisis.

11

Chapter 11: Economic and Social Implications

11.1 Introduction

The availability and access to clean water have profound economic and social implications for individuals and communities in developing countries. In this chapter, we explore the broader impact of water access on the economy, society, and individuals.

11.2 Economic Impact of Water Access

The economic implications of water access are multifaceted:

- Agriculture: Clean water is essential for crop irrigation and livestock, impacting food production and agricultural livelihoods.

- Productivity: Adequate access to clean water enhances productivity, leading to economic growth in various sectors.

- Healthcare Costs: Access to clean water reduces healthcare expenditures by preventing waterborne diseases and associated treatment costs.

- Gender Empowerment: Women and girls who spend less time collecting water have more opportunities for education and income-generating activities.

11.3 Social and Gender Implications

The social implications of water access extend to various aspects of life:

- Health and Hygiene: Clean water access leads to improved health and hygiene practices, reducing illness and mortality rates.

- Education: Adequate water supply in schools enhances attendance, particularly among girls who are less likely to miss school for water-related tasks.

- Women's Empowerment: Access to clean water can reduce the burden on women and allow them to participate in decision-making and income-generating activities.

11.4 Water as a Human Right

Water is not just an economic or social resource; it is a fundamental human right. Recognizing water as a human right is a significant step toward ensuring equitable access to clean water.

11.5 Inequality and Disparities

Despite progress, inequalities persist in water access, with marginalized communities often left behind. This chapter explores the disparities in water access and the need for equitable solutions.

11.6 Sustainable Development Goals (SDGs)

The United Nations' Sustainable Development Goals include specific targets related to water and sanitation, emphasizing the importance of water access for achieving broader development objectives.

11.7 Conclusion

Access to clean water is intricately linked to economic development, social well-being, and the empowerment of individuals and communities. It is not merely an infrastructure challenge; it is a matter of human rights and equity. In the following chapter, we will discuss the path forward and the importance of collective action to address the global water crisis. Understanding the economic and social implications of water access is essential for building a world where everyone can enjoy this basic human right.

12

Chapter 12: Future Prospects and Recommendations

12.1 Introduction

In this final chapter, we look toward the future and consider the path forward to address the ongoing global water crisis. We discuss recommendations and actions required to ensure that clean water becomes a fundamental human right for all, with a focus on sustainable and equitable solutions.

12.2 The Urgency of Action

The challenges related to water access in developing countries are pressing. Populations are growing, climate change is affecting water sources, and the consequences of inadequate access continue to harm communities. The urgency of taking action cannot be overstated.

12.3 Policy Recommendations

Effective policies are essential for addressing the global water crisis:

- Recognition of Water as a Human Right: Governments and international bodies should formally recognize the human right to clean water and sanitation.

- Investment in Infrastructure: Governments and organizations should invest in building and maintaining water infrastructure and distribution networks.

- Water Quality Regulation: Robust regulations should be in place to ensure the quality and safety of drinking water.

- Community Empowerment: Encourage and support community-led initiatives that empower local residents to manage and maintain water resources.

12.4 International Collaboration

Cooperation on a global scale is critical:

- International Aid: Donor countries and organizations should continue to provide aid and assistance to developing countries for water projects.

- Research and Innovation: Promote research and innovation in water-related technologies to make water treatment and distribution more efficient and cost-effective.

12.5 Education and Awareness

Raising awareness about the importance of water access and hygiene practices is fundamental. Educational programs should be implemented to inform communities about water issues.

12.6 Sustainability and Equitability

Efforts to improve water access must be sustainable and equitable:

- Sustainable Solutions: Projects should be designed with long-term sustainability in mind, with community participation in management.

- Equitable Access: Solutions should prioritize equitable access for marginalized communities and address disparities in water access.

12.7 The Role of Individuals

Individuals can also contribute to addressing the water crisis:

- Water Conservation: Implement water-saving practices in daily life to reduce water wastage.

- Supporting Initiatives: Donate to or volunteer with organizations working to improve water access in developing countries.

12.8 The Path Forward

The path forward to ensure clean water for all is one of collective action. It requires the efforts of governments, organizations, communities, and individuals. The global water crisis is a complex challenge, but by working together and taking action, we can make significant progress.

12.9 Conclusion

Access to clean and safe drinking water is a fundamental human right that impacts health, well-being, and development. The global water crisis is a challenge that can be overcome with the right policies, collaboration, innovation, and awareness. It is our collective responsibility to ensure that future generations have the basic right to clean water.